Ángeles García Aliaga

Hereditary Breast and Ovarian Cancer Syndrome

Ángeles García Aliaga

Hereditary Breast and Ovarian Cancer Syndrome

Cross-sectional study in patients at high or
moderate risk of SCMOH syndrome.

ScienciaScripts

Imprint

Any brand names and product names mentioned in this book are subject to trademark, brand or patent protection and are trademarks or registered trademarks of their respective holders. The use of brand names, product names, common names, trade names, product descriptions etc. even without a particular marking in this work is in no way to be construed to mean that such names may be regarded as unrestricted in respect of trademark and brand protection legislation and could thus be used by anyone.

Cover image: www.ingimage.com

This book is a translation from the original published under ISBN 978-3-639-55776-3.

Publisher:
Sciencia Scripts
is a trademark of
Dodo Books Indian Ocean Ltd. and OmniScriptum S.R.L publishing group

120 High Road, East Finchley, London, N2 9ED, United Kingdom
Str. Armeneasca 28/1, office 1, Chisinau MD-2012, Republic of Moldova, Europe
Printed at: see last page
ISBN: 978-620-7-65698-1

SUMMARY

Given the high percentage of breast cancer cases with high familial aggregation that are not explained by alterations in the main susceptibility genes, *BRCA1* and *BRCA2*, we proposed this study with the aim of determining the mutational prevalence of the *RAD51C* gene and the *RAD51D* gene in our population. These results will be useful for the assessment of *RAD51C* and *RAD51D* as risk predictors of hereditary breast and ovarian cancer syndrome (HBCS). For this purpose, a cross-sectional study was performed on patients meeting criteria for high or moderate risk of SCMOH syndrome in whom no pathogenic gene variants in *BRCA1*, *BRCA2* and *CHEK2 were* found to be associated with increased risk. Mutational screening of the index cases was performed by automated sequencing of the study areas (polymerase chain reaction of the amplicons, amplification check by agarose gel, enzymatic purification of the amplicons, sequencing reaction, column purification and sequencing by capillary electrophoresis). The sequences obtained were analyzed using the programs "SeqScape" v.2.5™ and "Sequencing Analysis" v.5.2. For the analysis of the variants, the main databases were consulted and bioinformatics and family cosegregation studies were carried out to evaluate their pathogenicity. The results were first compared with publications related to the *RAD51C* and *RAD51D* genes that are related to the Spanish population and that provide us with information on the variants found as well as the necessary primers and their conditions. The laboratory molecularly characterized the pathological gene variants and variants of uncertain clinical significance (VSCI) in order to determine the mutational profile of our population, calculating the prevalence of these variants. We also performed an analysis by clinical subgroups (bilateral breast cancer / ovarian cancer / breast and ovarian cancer), in order to perform a genotype-phenotype correlation study.

Key words: Molecular Genetics, Hereditary Breast and Ovarian Cancer Syndrome (HBOS), *BRCAs, BRCAX, RAD51C, RAD51D.*

ABSTRACT

Given the high percentage of breast cancer cases with high familial aggregation that are not explained by alterations in the major susceptibility genes, *BRCA1* and *BRCA2*, we propose this study to determine the prevalence of *RAD51C* and *RAD51D* genes mutation in our population. These results will be useful for the assessment of *RAD51C* as a predictor of cancer risk in hereditary breast and ovarian cancer syndrome (HBOCS). To do this, there has been a cross-sectional study on patients who meet criteria for high or moderate risk of HBOC syndrome and in which no gene variant in *BRCA1, BRCA2* and *CHEK2* associated with increased risk has been found. Mutation screening of index cases were made by automated sequencing study areas (Polymerase chain reaction, amplification checking by agarose gel, enzymatic purification of the amplicons, sequencing PCR, purification columns and capillary electrophoresis). The sequences obtained were analyzed using the software "SeqScape v2.5™" and "Sequencing Analysis 5.2". For the analysis of the variants will consult major databases, bioinformatics studies will be conducted and family cosegregation to asses pathogenicity of them. The results were compared in first place with related publications of *RAD51C* and *RAD51D* genes that are related to the Spanish population and provide us information on the variants we found, also necessary primers and conditions. From our laboratory were molecularly characterized the pathological gene variants and uncertain clinical significance variants (UCSV) in order to determine the mutation profile of our population, calculating the prevalence of these. Also there will be a clinical subgroup analysis (bilateral breast cancer/ ovarian cancer / breast and ovarian cancer), in order to conduct a genotype-phenotype study.

Keywords: Molecular Genetics, Hereditary Breast and Ovarian Cancer Syndrome (HBOCS), *BRCA's, BRCAX,* RAD51C, *RAD51D.*

INDEX

I. INTRODUCTION

Hereditary Breast and Ovarian Cancer Syndrome (HBOHCS)

Hereditary breast and ovarian cancer syndrome (HBOCS) is an inherited tendency to get breast, ovarian and related minority cancers. Although most cancers are not inherited, about 5% of people who have breast cancer and about 10% of women who have ovarian cancer have HCOHS.

Multiple environmental, hormonal and genetic risk factors have been identified in the development of hereditary breast and ovarian cancer syndrome. Although most of them have a slight influence, age, breast density, previous breast history and the number of first-degree relatives with breast cancer (BC) are particularly important in premenopausal women. In ovarian cancer (OC) the risk rises with age (with an increase in incidence at age 50) and decreases inversely with full-term pregnancy [1].

The understanding of the genetic cause of breast cancer gave way to the identification in the 1990s of the *BRCA1* and *BRCA2* ("breast-cancer-susceptibility") genes closely related to breast cancer susceptibility and first genes associated with SCMOH [2;3].

These genes are characterized by autosomal dominant inheritance, high penetrance and high prevalence. Mutations in *BRCA1* or *BRCA2* are not only associated with an increased risk of breast cancer, but also increase susceptibility to ovarian cancers in women and prostate, pancreatic and breast cancers in men [4].

But as in other hereditary cancer syndromes, there is a percentage of cases that are not explained by mutations in known high penetrance genes. SCMOH is one of the most striking, as it is generally considered that only 20-25% of high- and moderate-risk cases are explained by mutations in *BRCA1* or *BRCA2* [5]. This means that in most cases, the strong familial aggregation observed has no known genetic cause.

Prevalence and current status

Breast cancer is the most frequent cancer and one of the main causes of mortality in women in the western world, since each year 1.38 million new cases are diagnosed and it causes 458,000 deaths worldwide. In European Union countries it is estimated that the probability of developing this disease is 8% before the age of 75 years [6]. In Spain, the incidence of breast cancer is one of the lowest in Europe, even so it is estimated that 22,000 cases of breast cancer are diagnosed each year, accounting for 30% of all tumors diagnosed in women. As for the region of Murcia, 560 women are diagnosed annually with invasive breast cancer and about 180 die from this cause.

On the other hand, more than 150,000 cases of ovarian cancer (OC), the gynecologic cancer with the highest mortality, are diagnosed worldwide each year, as it is usually diagnosed in advanced stages [7].

As discussed, one of the main risk factors is the presence of breast cancer in immediate family members. The risk of MC doubles in first-degree relatives of women with MC, while OC triples with relatives affected with this disease compared to women with no family history [8].

Of all cases with a hereditary component, 5-10% are attributable to autosomal dominantly inherited mutations in several susceptibility genes. An additional 15% of women with breast cancer have some family history, although without a clear pattern of inheritance [9].

Molecular Genetics

Linkage analysis of families with multiple cases of breast cancer and ovarian cancer identified in 1990 the first chromosomal region (17q21) candidate to contain a high-risk gene for hereditary breast cancer. [10]. In 1994, by positional cloning, the causal gene, named *BRCA1* ("BReast CAncer gene 1"), was identified [MIM *113705] [11]. *BRCA1* is a gene consisting of 24 exons, generates several transcripts, and the most important is the one encoding the largest protein of 1863 amino acids.

The functions exerted by the *BRCA1* protein through the complexes it forms are involved in the activation of responses to DNA damage and in cell cycle control.

In 1995 *BRCA2* ("BReast CAncer gene 2") [MIM *600185], the second high penetrance gene implicated in hereditary breast cancer, was identified [12] [12]. [12]. *BRCA2* is a larger gene than *BRCA1*, is located on chromosome 13q12.3 and has 27 exons, encodes a protein of 3418 amino acids, which has fewer identified structural motifs.

The main function of *BRCA2* is homologous recombination (HR), which is based on its ability to bind to the *RAD51* strand invasion recombinase.

Although the possibility of identifying a new allele with high penetrance is not ruled out, it is currently suggested that multiple susceptibility genes are involved, which are distributed into three groups depending on the relative risk to which they are related. The groups into which these genes are rearranged are:

-**High penetrance** genes: Associated with an increased relative risk (RR) of MC higher than 5, the *BRCA1* and *BRCA2* genes are the ones that confer the highest number of molecular of molecular characterizations of SCMOH with 25% of the diagnoses of this syndrome and other minority but high penetrance genes such as TP53, PTEN, STK11 and CDH1 that cause 5% of familial CM.

-**Moderate penetrance** genes: confer a cancer RR between 1.5 and 5.   related to Fanconi Anemia (FA) such as *BRIP1, PALB2,   RAD51C* and *XRCC2* and other genes that are not involved in FA, such as   *ATM, CHEK2, NBS1, RAD50, RAD51D* and *RAD51B*. All these genes   would account for   5% of molecular diagnoses of SCMOH.

-**Low penetrance** genes: They have an RR of around 1.5 and are a group of at least 67 genes currently identified that could explain 14% of CM. currently identified, which could explain 14% of familial CM. familial.

Compared to the risk of *BRCA1* and *BRCA2*, the combination of these secondary genes accounts for approximately 10% of the genetic component of CM risk [8]. In these categories of susceptibility alleles where variants with moderate and low penetrance are included we should highlight the *ATM, CHEK2, PALB2* and *BRIP1* genes, as their proteins interact with *BRCA1* and *BRCA2* proteins or due to their involvement in the same DNA repair pathways.

However, about 51% of patients with familial breast cancer who do not show mutation in these genes fall into the *BRCAX* family category. These families may either carry a mutation in an unidentified moderate-penetrance gene or a polygenic pattern involving several low-penetrance loci.

There are also other genetic models that can explain the familial risk of CM not due to *BRCA1* or *BRCA2*. Other genes with high penetrance mutations have been identified in hereditary syndromes that include CM as part of the phenotype (such as the *TP53* gene responsible for the P53 protein in Li-Fraumeni syndrome or the *PTEN* gene that transcribes a phosphatase enzyme in Cowden syndrome) [1314] but as these are low prevalence alleles they are not considered causative of SCMOH.

Overall, the two most important gene clusters comprise less than 35% of the familial risk of CM [14;15]. Therefore, the cause of most familial clustering remains unknown in at least 51% of cases, which may be due to a large number of variants, each associated with a moderate effect on CM risk (**Figure 1**).

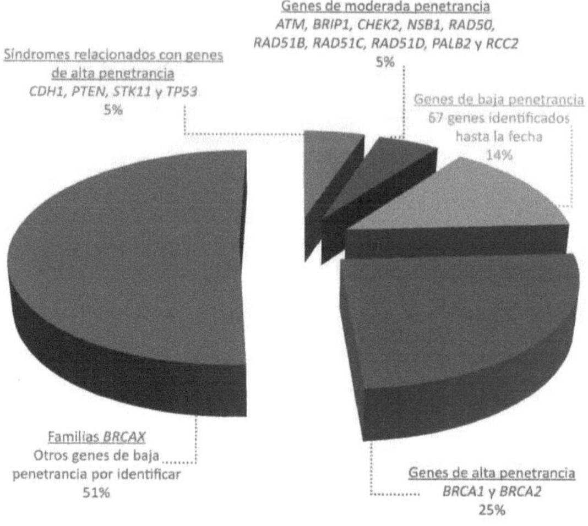

Figure 1*Breast cancer susceptibility genes.*

In this regard, an important study was published in 2010 by Meindl et al. in which *RAD51C* analysis was performed by sequencing and dHPLC (denaturing high-performance liquid chromatography) in 1,100 probands from German families with CM/CO without mutations detected in *BRCA1* or *BRCA2* (*BRCAX*). As results, six monoallelic mutations conferring increased risk for CM and CO (two insertions, two splicing site mutations and two missense mutations) were identified as limiting protein function. A mutational prevalence in the *RAD51C* gene of 1.3% was evidenced if families with individuals affected by breast and ovarian cancer were selected, and families with individuals affected by breast cancer alone were discarded [116]. At the same time, a recent study carried out in Spanish families showed a prevalence similar to that described in the previous article for the mutations found in *RAD51C* [17]. With respect to the *RAD51D* gene, an important study in a German population assigns a relative risk of ovarian cancer of 6.3 in patients carrying pathogenic gene variants present in this gene, suggesting that the genetic study of this gene can be of great clinical utility in families with SCMOH in which there are individuals affected by ovarian cancer [18].

Molecular Basis

In addition to *BRCA1* and *BRCA2*, other CM susceptibility genes (*ATM, CHEK2, BRIP1 or PALB2*) are essential for cellular genomic stability and have been functionally related to DNA repair.

Homologous recombination is a fundamental process for the conservation of organisms as it maintains genomic integrity by repairing DNA double strand breaks (DSBs). This damage is generated during DNA replication (S phase of the cell cycle) and preferentially repaired by recombination with the sister chromatid. A dysfunction in HR can cause aberrant gene rearrangement and genomic instability, leading to chromosomal translocations, deletions, amplifications and loss of heterozygosity (LOH), a phenomenon by which a "locus" (precise position of a gene on a chromosome) loses one of the copies of a gene.

The importance of the correct functioning of the HR process is evident in pathologies such as Fanconi anemia and other neoplastic syndromes that share chromosomal instability in childhood, developmental abnormalities and predisposition to leukemias and other types of cancer. In these syndromes bi-allelic mutations have been observed in the *BRCA2, PALB2 and BRIP1* genes. This evidence suggests that, in the case of the *BRCA2* gene, when two alleles with the same mutation are present, it causes FA; however, if only one of them is mutated, it is related to CM.

HR appears to be more important in correcting replication errors than exogenous DNA damage in DSB, in which case other repair pathways could be used (**Figure 2**). [19]. Similarly, tumor suppression would be intimately related to replication error correction rather than DSB repair since if damage is not resolved, cells may remain arrested until an apoptotic pathway is activated. In the absence of such control responses, sufficient time may not be allowed for the DSB to be fully repaired before entry into mitosis and may lead to genome instability, manifested by chromosomal loss and/or missegregation.

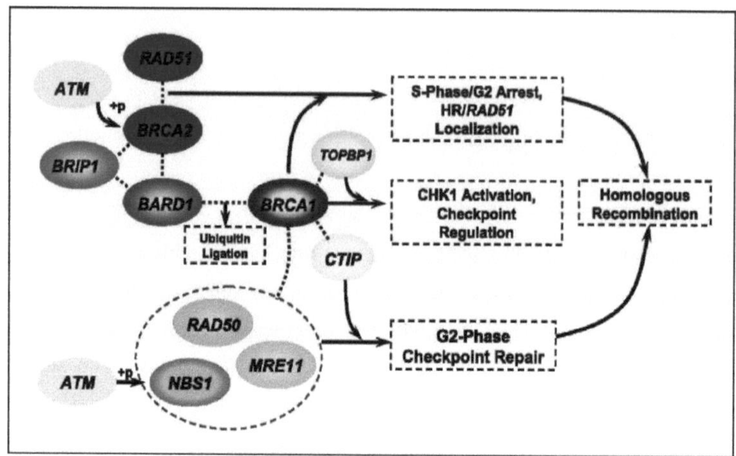

Figure 2: *Homologous Recombination* (13).

BRCA1 and *BRCA2* expression under normal conditions is related to the growth and differentiation of mammary epithelial cells, especially during gestation and lactation. Although these genes are ubiquitously expressed, it appears that their activity has a particular impact on breast cells [20].

These genes act in DNA repair (prevent neoplastic transformation), and follow an autosomal dominant pattern of inheritance with high penetrance. Carriers of mutations in any of these genes have a high risk of developing breast and/or ovarian cancer during their lifetime. Therefore, the detection of these mutations allows the localization of asymptomatic carriers who have a high lifetime risk of developing cancer.

BRCA1 harbors a highly conserved amino-terminal region (*RING* domain), a coiled-coil domain and tandem domains in the carboxyl-terminal region (*BRCT* domains). The *RING* domain is a motif in which numerous ubiquitin (E3) ligases that mediate protein ubiquitination are found. The sequences spanning this domain mediate a stable association with *BARD1* ("BRCA1-Associated RING domain protein 1") [21]. *BRCA1/BARD1* dimer formation is implicated in the maintenance of genomic stability and tumor suppression through its involvement in DNA damage signaling, DNA repair, and transcriptional regulation [22].

CTIP ubiquitination by *BRCA1* is involved in DSB repair through its interaction with the *MRN* complex (formed by *MRE11, RAD50* and *NBS1*). The N-terminal domain of *BRCA1* interacts with another similar *RING* domain called *BARD1* and thus they form an E3 ligase heterodimer. The interaction of *BARD1* is required to stabilize the conformation of the *RING* domain of *BRCA1* for E3 activity. This ubiquitination stabilizes *BRCA1* and increases its E3 activity, involved in DSB repair and chromatin remodeling.

The involvement of *BRCA1* has also been shown, which phosphorylates at the late stage of HR and recruits the *RAD51* protein to the damaged DNA loci through its interaction with BRCA2 [23]. [23]. The *BRCA1-BARD1* complex is involved in the activation of G1/S checkpoints (through phosphorylation of *BRCA1* via *ATM* or *ATR*), S phase and G2/M [24]. The *TOBP1* complex ("*BRCA1-BRIP1-DNA* topoisomerase II-binding protein 1") is required in the control of S-phase and DNA replication [25].

The *BRCA1* protein is capable of forming numerous complexes, each of which acts in the activation of the DNA damage response, in the control of cell cycle activation and/or in the repair of double-strand breaks.

On the other hand we have the *BRCA2* protein which contains eight *BRCT* repeats, each of which can bind to *RAD51*, and a DNA binding domain. *BRCA2* mediates the recruitment of *RAD51* recombinase to DSBs and is responsible for the tumor suppressor function of this repair process. [26]. In addition, the C-terminal end of *BRCA2* contains a cyclin-dependent kinase (*CDK*), which also binds RAD51 [27].

RAD51 has been shown to catalyze the final biochemical step in HR: the exchange of strands [28]during which single-stranded DNA (ssDNA) invades the homologous duplex DNA, displacing the identical strand of the duplex and forming a displacement loop.

Recently, a homozygous missense mutation in the *RAD51C* gene (17q22-q23) was identified in a family with PA characteristics and high cellular susceptibility to agents causing DNA lesions. This gene of the *RAD51 family* is involved in DNA repair by homologous recombination. The protein accumulates at sites of DNA damage together with *RAD51*, acts in some phases of the repair process and is also involved in *CHEK2* activation and cell cycle arrest in response to DNA damage. Therefore, *RAD51C* contributes to preserving genome integrity [29].

II. HYPOTHESIS

From April 2007 to the present day, the Hereditary Cancer Genetic Counseling Clinic of the Oncology Department of the Hospital Clínico Universitario Virgen de la Arrixaca (HCUVA) has been assessing several hereditary cancer syndromes, including hereditary breast and ovarian cancer syndrome. Of all the families referred, and under selection criteria of high risk for this syndrome, a genetic study of the two genes recommended by the clinical guidelines (*BRCA1* and *BRCA2*) has been requested in a total of approximately 600 index cases, study performed in the Genetic Diagnostic Laboratory (LDG) of the Clinical Analysis Service of the HCUVA. Taking into account that pathogenic gene variants have been found in about 20% of the families studied, there is still a high percentage of families that have not been genetically characterized.

Currently, there are several published studies that also describe an increased risk of breast and ovarian cancer in patients carrying gene variants present in other susceptibility genes. However, the distribution and frequency of these variants can be very different depending on the geographical area analyzed. In the Region of Murcia, we have an important registry, both of pathogenic variants, variants of unknown clinical significance and non-pathogenic variants obtained from genetic studies performed in selected families, where the presence of founder mutations has already been described [30]. The study of other hereditary breast and ovarian cancer susceptibility genes is of vital importance to establish relationships between the gene variants obtained and the phenotype of the disease (genotype-phenotype correlation), as it is a tool of enormous importance in genetic counseling.

Thus, in the case of finding pathogenic variants in any of the genes in the study, with a prevalence similar to that described in the literature, the genetic study of that particular gene will be incorporated into the portfolio of services of the HCUVA hospital, in order to increase the performance of the genetic tests performed. In this way, in addition to the index cases, a large number of family members will be able to benefit from the results obtained after performing these additional genetic tests.

III. OBJECTIVES

1. Mutational screening of *RAD51C* and *RAD51D* genes in patients from the Region of Murcia meeting the selection criteria for hereditary breast and ovarian cancer syndrome, referred to the Genetic Counseling Unit of the Hospital Universitario Virgen de la Arrixaca during the years 2007 and 2014, in which no pathogenic variants in *BRCA1* and *BRCA2* genes, nor any major gene rearrangements were found.
2. Classification of the variants found in the *RAD51C* and *RAD51D* genes according to their type and pathophysiological implications according to the effect they cause.
3. Study of variants of unknown effect detected by bioinformatics studies.
4. Study of the genotype-phenotype correlation in pathogenic variants and variants of uncertain significance found in the genetic study.
5. Recruitment of healthy controls for the assessment of gene variants found during the study.

IV. METHODOLOGY

1. Patient recruitment

From the cohort of approximately 600 index cases belonging to families meeting high-risk criteria for hereditary breast and ovarian cancer syndrome, those cases that had completed the genetic study of the *BRCA1/2* genes, that were not carriers of pathogenic genetic variants in these genes and that were negative in the analysis of large gene rearrangements were selected by the physicians who make up the Genetic Counseling Consultation. Of these, high-risk families that met the inclusion criteria considered by the members of the Genetic Counseling Consultation, where the specific bibliographic recommendations for each of the genes were taken into account, were recruited for the present study. The molecular study of the candidate genes *RAD51C* or *RAD51D* was extended to these patients.

Before the test was performed, patients were informed, by informed consent, of the objective of the study and the implications and limitations of genetic testing (**APPENDIX 1**).

2. Genetic study

2.1 <u>The genetic study of the *RAD51C* and *RAD51D* genes of the selected patients</u>, as well as the interpretation of the gene variants obtained was carried out by the physicians of the LDG. Using the genomic library available in the LDG of the Clinical Analysis Service of the HCUVA, the DNA of the selected samples will be used to amplify by PCR all the exons of both genes and their exon-intron boundaries.

2.2 <u>Clinical characterization of the genetic variants found by searching databases</u>: The clinical relevance of the gene variants in the study was consulted in current publications and in the different available databases: 'Human Gene Mutation Database' (HGMD), 'Leiden Open Variation Database' (LOVD), 'dbSNP' (NCBI) and 'Ensembl'. Based on the information obtained, the variants were classified as: non-pathogenic, of unknown or uncertain clinical significance (VSD/VSCI) and pathogenic.

2.3 <u>Familial cosegregation study in the case of pathogenic gene variants and ICHV</u>: The nursing staff of the Hereditary Cancer Genetic Counseling Clinic contacted by telephone the relatives of the index case under study, who will be referred to the clinic, and after receiving genetic counseling by the corresponding oncologist, a peripheral blood collection was performed. After this, the samples were sent to the LDG, to be processed by automated DNA extraction.

2.3 <u>Assessment of the clinical relevance of variants of unknown clinical significance</u>: If the variant obtained in the study was not published or registered in any of the databases consulted, a predictive bioinformatics study of its possible pathogenicity was carried out, as well as the estimation of the allelic frequency for the change detected in the population of healthy controls. For the bioinformatic studies, the available programs were used according to the changes found:

-Missense variants: Grantham Score Matrix (GMS), Align-GVGD, SIFT, Polyphen-2, pMut.

-Missense and silent variants: "MutationTaster"

-Effect of variants in splicing: Splice Site Finder, NNSplice and MaxEntScan. "MaxEntScan

3. Procedure

3.1 Genomic DNA Extraction

We worked from a peripheral blood sample in a tube with EDTA anticoagulant. Genomic DNA (deoxyribonucleic acid) from all index cases was extracted from 400□L using Promega's automatic system (Maxwell 16 Blood DNA Purification Kit. **Figure 3**). This technique is based on the interaction of paramagnetic particles that function as a mobile solid phase that optimizes the uptake, washing and elution of the DNA present in the leukocytes of the sample.

Figure 3. Maxwell 16 System.

After extraction, DNA concentration and purity were measured spectrophotometrically using Thermo Scientific Nanodrop 1000 equipment (**Figure 4**). The measurements were performed at a wavelength of 260 nm, at which nucleic acids absorb, informing us of the DNA concentration in ng/$\square$L. The ratio obtained between the absorbances at A_{260}/A_{280} nm determines the quality of the extracted DNA, a ratio between 1.5-1.8 being considered acceptable.

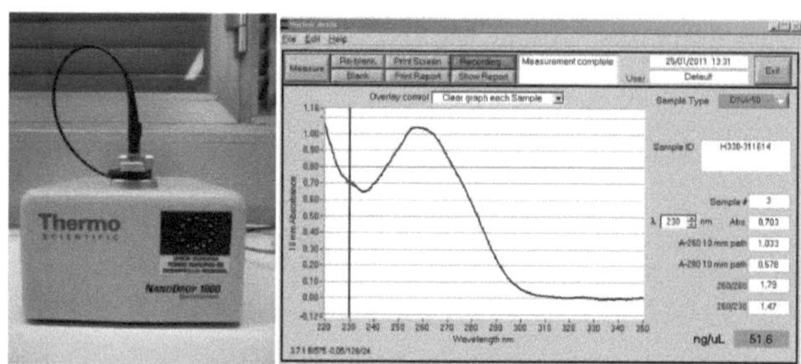

Figure 4. (A). Absorbance curve of a DNA sample measured on the Nanodrop1000 (Thermo.) (B). Nanodrop1000 (Thermo)

3.2 Amplification and purification of genomic DNA

For the amplification it was necessary to use primers flanking the exonic zones to be studied, for this it was necessary to design these primers according to the complementary target sequence that we wanted to amplify. To obtain double the yield we used two primers, one in the "*Forward*" direction and the other in the "*Reverse*" direction of the complementary chain, both in the 5'$\rightarrow$ 3' direction of the chain.

The primers used for the amplifications are detailed in **Table 1** for the *RAD51C* gene and in **Table 2** for the *RAD51D* gene. The average melting temperature (T^am) recommended by the manufacturer or experimentally determined in the laboratory and the size of the resulting amplicons are also indicated.

We used as reference sequence the isoform A, the longest isoform of the *RAD51C* gene (NM_058216.1, OMIM *602724). Each of the 9 exons of the *RAD51C* gene was amplified using the Polymerase Chain Reaction (PCR) technique.

Exon	T^a_m	Primer	Amplicon size (bp)
1	58º	F 5'-T AGCAGAATCTAACGGAGACT-3' F 5'-T AGCAGAATCTAACGGAGACT-3' R 5'-ACAAGACTGCGCAAAGCTG-3' R 5'-ACAAGACTGCGCAAAGCTG-3'	294
2	58º	F 5'- TCCACTCCTAGCATCATCACTGTT -3' R 5'- CCCACCCTTAAAAGGAGAAC -3'	399
3	60º	F 5'-TCATGATTTGGTTTTGTTTGTTTGTCATC-3' R 5'- GGTCTCAGATGGGCACAAAT-3' R 5'- GGTCTCAGATGGGCACAAAT-3'	274
4	60º	F 5'- TGCCAATACATCATCCAAACAGG-3' R 5'- CAGGCAAACGCTATTTTGAC-3' R 5'- CAGGCAAACGCTATTTTGAC-3'	249
5	60º	F 5'- TCTTGGAGAGAGAGAGAGAGAGAGAGCATTTT-3' R 5'-CAGGCAAACGCTATTTTGAC-3' R 5'- CAGGCAAACGCTATTTTGAC-3'	299
6	62º	F 5'- TGGGGGGTTTTTCACAATCTTGG-3' R 5'- GTCTGCTTTCATCATGAAGCGTATATAGT-3'	251
7	60º	F 5'- TCTTGGAGAGAGAGAGAGAGAGAGAGCATTTT-3' R 5'- GTCTGCTTTCATCATGAAGCGTATATAGT-3'	243
8	62º	F 5'- ACGGGTAATTTGAAGGGTGTGT-3' F 5'- ACGGGTAATTTGAAGGGTGT-3' R 5'- AGCATCAAAAGCTGTCCTCA-3' R 5'- AGCATCAAAAGCTGTCCTCA-3'	362
9	62º	F 5'- GCCTGGGGCCCCCTAGAATAAAGT-3' R 5'- GGTATTTTTTTTTCCCATTCATTCACTTCA3-3'	341

For the other gene we also used as reference sequence the longest isoform A of the *RAD51D* gene (NM_001142571.1, OMIM *602954). Each of the 10 exons of the *RAD51D* gene is amplified by PCR.

Exon	T^a m	Primer	Amplicon size (bp)
1	62º	F 5'-TGTCCTCCTCTCTAGGAAGGGGTA-3' R 5'-AGGTATGCCAGGGCAGTG-3'	345
2	60º	F 5'- AGGCCCCAGTCCTCTCTCTCTG -3' R 5'- AGACACTCAGGTTTGGAATGTG -3' R 5'- AGACACACTCAGGTTTGGAATGTG -3'	230
3	58º	F 5'- TTTGTTGCTGAGCAGTCTGG-3' F 5'- TTTGTTGCTGAGCAGTCTGG-3' R 5'- CCTCCTGCCTCTCTCTCTCTCCTCTTCT- 3'	317
4	58º	F 5'- TTTTCCCCTGTCTTCTTCCTCCTCCT- 3' R 5'- ACCACCCTCCTCACCCACCCCTAAATC-3'	263
5	58º	F 5'- CCATCTGGACCCTCTCTCTCTGAA-3' R 5'- GGGGTTTTCCTGTGTCAGAA-3' R 5'- GGGGTTTTCCTGTGTCAGAA-3'	305
6	58º	F 5'- CCCCCCCTACTCCCTCTCTCTCTTATCC-3' R 5'- AGTAGGACACCTGCCCACACAG-3' R 5'- AGTAGGACACCTGCCCACAG-3'	225
7	58º	F 5'- GCTGACACAGGTTCATGAGTGC-3' F 5'- GCTGACAGGTTCATGAGTGC-3' R 5'- GCCAGAGAGAGACCAGACTCCAGA- 3'	234
8	58º	F 5'- CCTCCCTCTCTCTGCTTCTCCT-3' F 5'- CCTCCCTCTCTCTGCTTCTCCT-3' R 5'- TTTGGGGGGGTTCAGAAGCTGAC-3' R 5'- TTTGGGGGTTCAGAAGCTGAC-3'	324
9	58º	F 5'- AGCATTATGGATCTGTAAGTCTG-3' -	223

		F 5'- AGCATTATGGATCTGTAAGTCTG-3' R 5'- CCTCCAGGGGGCCCCCAAGATTA-3'	
10	58°	F 5'- GAGGCTGAAACCTTGCAACT-3' F 5'- GAGGCTGAAACCTTGCAACT-3' R 5'- CAGGCGTTACTGGGAAGAAGAAA-3'	350

PCR is a molecular biology technique whose objective is to obtain a large number of copies of a DNA fragment of a certain exon of the gene we are studying from a specific patient, starting from a minimum concentration of DNA.

This technique is used to amplify a DNA fragment, in this case each of the exons of the *RAD51C* and *RAD51D* genes separately in order to obtain their sequence.

The amplification reactions were performed in a final volume of 25☐L, the components of the PCR reaction, their concentrations and the volume used of each is determined in **Table 3.**

Componentes PCR	V (µl)	V (µl) X 5	[Final]
Agua	X (10.88)	54.4	-
Buffer (5 X)	5	25	1 X
Cl2Mg (25 mM)	1 (2)	10	2 mM
Primer Forward (10µM)	1	5	0.4 µM
Primer Reverse (10 µM)	1	5	0.4 µM
DNTPs (2mM)	2.5	12.5	0.2 mM
Taq/Promega (5U/µl)	0.125	0.625	0.625 U
DNA genómico (20 ng/µl)	2.5	-	50 ng
Vf	25		

Table 2: RAD51D gene primers

Table 3. *Polymerase Chain Reaction*

The thermal cyclers used were the following: "2720 Thermal Cycler" from "Applied Biosystems", "GeneAmp PCR System 9700" and "Veriti" model from the same house. The enzyme kit chosen for the amplifications was the "Promega Go Taq Hot Start polymerase".

The Promega enzyme requires a shorter activation time than other commercially available enzymes and reduces the formation of nonspecific products and primer dimer formation.

PCR was performed in several steps within the thermal cycler:

-Polymerase reactivation phase at 94°C for 2 minutes.

PCR cycles
from 30-35 cycles

1st: Denaturation of the template strand at 94°C for 1 minute –
2nd : Annealing temperature that varies depending on the primer, ranging 58°C to 62°C as indicated in the table for 45 seconds.
3°: Final extension phase at 74°C for 7 minutes.

-Once the PCR cycles have been completed, the amplified products remain at 4°C (programmed as HOLD) until they are collected from the thermal cycler.

To verify the correct amplification of the fragments, electrophoresis was performed in 1.5% agarose gel with 1X TBE buffer (Tris 89mM - boric acid 89 mM-EDTA 2mM at pH8.4. "Bio-Rad") using for developing "GelRed" (0.1 □L ΓελPεδ/1μΛ gel) "GelRed Nucleic Acid Gel Satin, 10000X in Water", a fluorescent nucleic acid staining solution that replaces ethidium bromide (very toxic), commonly used in molecular biology laboratories for double-stranded DNA staining. Approximately 3□L of amplified product was loaded along with 1 □L of loading buffer 0.25% (W/V) bromophenol blue, 0.25% (W/V) cyanol xylene, 30% (V/V) glycerol in water. The size of the samples was compared with a molecular weight size marker "pGEM DNA Markers" from "Promega". The "Alpha Innotech" transilluminator and Canon "PowerShot A640 AiAF" camera were used for development.

The amplicons were then purified by an enzymatic method using the "Exosap-It" kit (**Figure 5**). This kit includes two enzymes, exonuclease I, which removes residual DNA single strands that may be formed in PCR or primer residues and an alkaline phosphatase that removes dNTPs residues. The PCR reaction product 5□L was mixed with 2□L of "Exosap-It", incubated 15 minutes at 37°C and then 15 minutes at 80°C for inactivation.

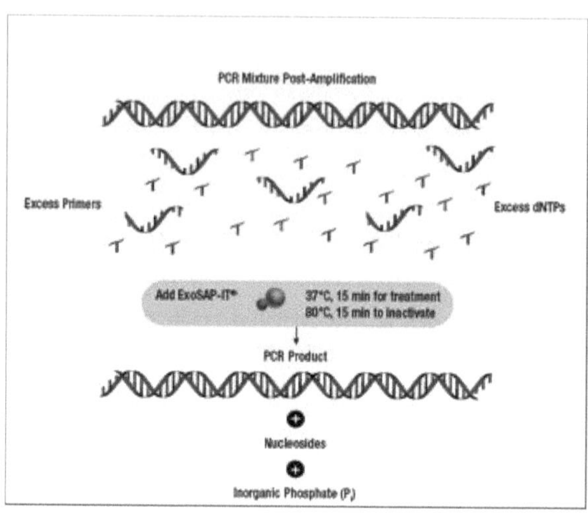

Figure 5: *Enzymatic reaction for PCR product purification.*

3.3 Sequencing Reaction

By bidirectional sequencing we were able to detect alterations in the gene sequence. A new PCR was performed with the purified amplicon using the "BigDye Terminator" (BDt) v1.1 kit from Applied Biosystems, an adaptation of the dideoxy enzymatic reaction of Sanger (1977). The sequences were analyzed by capillary electrophoresis on the "ABI3130", a four-capillary analyzer also from Applied Biosystems.

For the sequencing reaction, the same primers were used as for PCR but at a lower concentration of 3.2 µM. We worked with a final volume of 5 µL; 1.75 µL "Buffer enhancer sequencing" (10X), 1.5 µL miliQ water, 0.25 µL of BDt v1.1, 0.5 µL primer ("Forward" or "Reverse" 3.2 µM) and finally 1 µL of the PCR product (**Table 4**).

PCR sequencing	**Vol (µL)**
H O$_2$	1,5
Buffer enhancer sequencing" (10X)	1,5
BigDye	0,5
Primer F or R	0,5
Purified amplicon	1
Total Vol.	5 µL

Table 4: Sequencing Reaction

The sequencing reaction or sequencing PCR follows the following scheme (**Figure 5**):

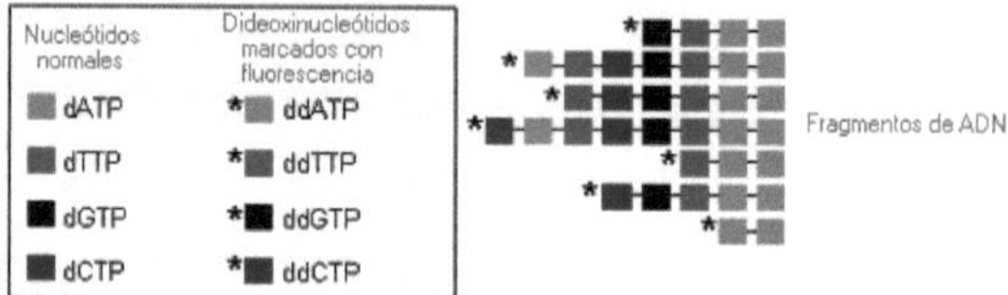

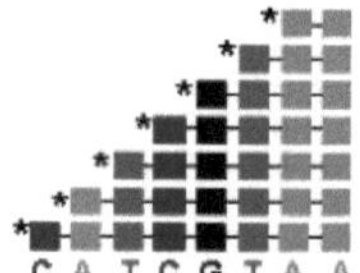

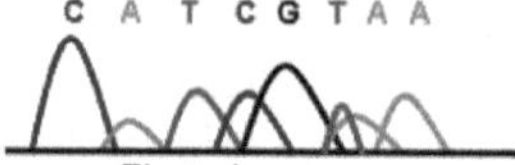

Longitud de onda	Nucleótido
Rojo	T
Verde	A
Azul	C
Negro	G

Figure 5: *Schematic of the sequencing reaction.*

-DNA strand denaturation phase at 96°C for 1 minute.

1st: Mold strand denaturation at 96°C for 1 minute.

2°: Annealing primer hybridization temperature at 50°C for 5 minutes.

-PCRsecond cycles

3°: Final extension phase at 60°C for 4 minutes.

3.4 Purification of the sequencing reaction

After the sequencing reaction, leftover dNTPs and possible impurities were removed with "EdgeBio" columns ("Performa DTR Dye Terminator Removal Gel Filtration Cartridges"). This kit is based on the Sambrook et al. (1989) description of DNA gel filtration to separate fragments larger than 16 base pairs (bp) from labeled dideoxynucleotide residues, dNTPs and other salts or low molecular weight compounds (removes up to 98% of salts present).

The protocol to be followed was as follows:

-The columns were centrifuged at 3000 rpm for 2 minutes.

-The remaining water was removed and the column was placed in a new 1.5 ml vial.

-10μL of milliQ water was added to the 5μL of sequencing reaction, the final volume (15μL) and deposited in the center of the column.

-Centrifuged again at 3000 rpm for 2 minutes.

10 μL of Formamide Hi-Di from Applied Biosystems was added to denature the DNA double strand, then the purified samples were transferred to a 96-well plate ("MicroAmpTM", "Optical 96-Well Reaction Plate", "Applied Biosystems") adapted for the sequencer.

3.5 Direct automatic sequencing by capillary electrophoresis

Capillary electrophoresis (CE) is a separation technique used to separate the different molecules present in a solution according to their mass/charge ratio. The fragments to be analyzed were attached to fluorescent tags, which were detected by a laser that excited them at different wavelengths, thus achieving the analysis of multiple fragments at the same time, with those of lower molecular weight traveling faster through the capillary and those of higher molecular weight traveling more slowly.

4. Data analysis

4.1 Alterations in the sequence

Sequence analysis was performed with the bioinformatics programs "SeqScape v2.5™" and "Sequencing Analysis 5.2". Within the sequence we were able to find different types of alterations that depending on their significance had more or less impact on the protein encoded by the *RAD51C* and *RAD51D* genes. For the interpretation of these results we took into account certain information:

Basic Information:

➢ <u>Type of mutation</u>
- Silent: nucleotide change without amino acid change, not usually pathogenic (except splicing sites).
- Mutations affecting mRNA maturation *("splicing")*: The result is the inclusion or exclusion of exons from the protein, depending on the point of splicing.
- Nonsense: *they* do not affect the coded amino acid sequence but generate a stop codon (UAA, UAG or UGA) that interrupts the translation of the protein prematurely.
- Missense: the substitution of a base gives rise to a new triplet or codon encoding a different amino acid.
- Deletions or insertions that alter the reading frame (*"frameshift"*): these are mutations in which the deletion or insertion of a nucleotide leads to a displacement of the triplet reading frame.
- Deletions or insertions that do not alter the reading frame: these are small deletions or insertions that are multiple of 3 and therefore introduce or delete one or a few amino acids, but do not change the reading frame of the triplets.

➢ <u>Region of the gene and protein affected</u>:

- Exons > flanking intronic region > intron
  - <u>Intronic region.</u> These gene variations may affect the recognition sites of the cleavage and splicing machinery, or may result in a possible deletion or appearance of the splicing acceptor or donor site.
- Effect dependent on the functional region of the <u>protein.</u>

➢ <u>Phylogenetic conservation of amino acids</u>:

- Changes in the most conserved areas between species or isoforms are usually more deleterious (pathogenic) than in other areas.

➢ <u>Bioinformatics studies</u>:

- *HGMD* (http://www.hgmd.cf.ac.uk/ac/index.php): Extensive private database on human inherited disease mutations in genetic and genomic research [32].

- *LOVD* (http://www.lovd.nl/3.0/home): This is a public database developed by "Leiden University Medical Center" in the Netherlands, designed to collect information on all types of variants of a large number of genes [33].

- *Ensembl Genome Browser* (http://ensembl.org/index.html): Bioinformatics research project for eukaryotic genomes and open access.

- *NCBI* (http://www.ncbi.nlm.nih.gov/): Its SNP database is very useful for the search of polymorphisms.

- When they are missense mutations there is a correlation between the physicochemical changes between amino acids and how they can affect the protein, in this case in-silico studies based on algorithms that predict the severity of the variant are very useful:
   - When dealing with intronic variants, analysis programs such as "MaxEnt" [34], "NNSPLICE" [35] and "HSF" [36] are used.
   - For the analysis of variants in coding regions we used the bioinformatics programs "SIFT" [37] and "Mutation Taster" [38] as predictors of pathogenicity.

- These are indicative and predictive studies, never sufficient to confirm or rule out pathogenicity.

➢ <u>Functional studies and animal models</u>: not very feasible in a clinical laboratory.

Clinical information:

➢ <u>Absence in healthy controls</u>: It is important to study at least 200 alleles of healthy individuals from the same population to validate the pathogenicity of the variants.

➢ <u>Cosegregation of the mutation with the disease</u>:
- To assess possible inheritance patterns and cases of incomplete penetrance.
- Difficulty in cases with several mutations (some affected individuals will have only one mutation).

➢ <u>Previous mutation descriptions</u>: information on penetrance, evolution and genotype-phenotype correlations.

4.2 Analysis of results and statistics

The mutations were analyzed in order to evaluate the mutational profile of our population, calculating the prevalence of these mutations.

An analysis by clinical subgroups (bilateral breast cancer / ovarian cancer / breast and ovarian cancer) was also performed, in order to achieve a genotype-phenotype study.

The study of families with gene variants was important to collect valuable information on the penetrance of the variant and consequently provide very useful information in risk prediction models.

V. DIFFICULTIES AND LIMITATIONS

The sample size needed to achieve all the objectives must be large, so the study can be extended over time.

The study covers a long period of time, and the immunohistochemical determinations on tumor tissue have varied over this period, so the immunohistochemical profile may provide different data depending on the date of the study.

The collection of clinical and demographic data of some patients can be difficult due to the diversity of origin of the patients. The patients in the genetic counseling consultation come from different hospitals in the region of Murcia.

VI. ETHICAL ASPECTS

The characteristics of our study design, descriptive cross-sectional, implied a clearly favorable risk-benefit balance for patients, since no high-risk intervention was performed on individuals.

All patients included in the study had to fill out an informed consent form, as well as the first-degree relatives to be studied.

VII. WORKFLOW

Stages of development

Step 1: Optimization of the conditions for the different molecular techniques.

Stage 2: Review and selection of retrospective cases for which a quality sample is available in the LDG files.

Stage 3: Recruitment of patients in the prospective section who met the inclusion criteria. These patients were informed at the Genetic Counseling Unit by the Physicians responsible for the Medical Oncology service. They themselves provided them with information on the ethical aspects and the impact of genetic testing. Before blood collection, the patient must sign the informed consent form.

Stage 4: Receipt of samples in the LDG and DNA extraction for the molecular techniques described in the methods section of this work. **Figure 6 shows** the flow diagram for sample processing.

Stage 5: Review of the available clinical information of the retrospective cases and those included prospectively during the development of the project (age at disease diagnosis, family history, immunohistochemistry, etc.). The data collected were included in a database generated specifically for this project.

Stage 6: Data processing and analysis.

Stage 7: Preparation of the final report and publication of the results.

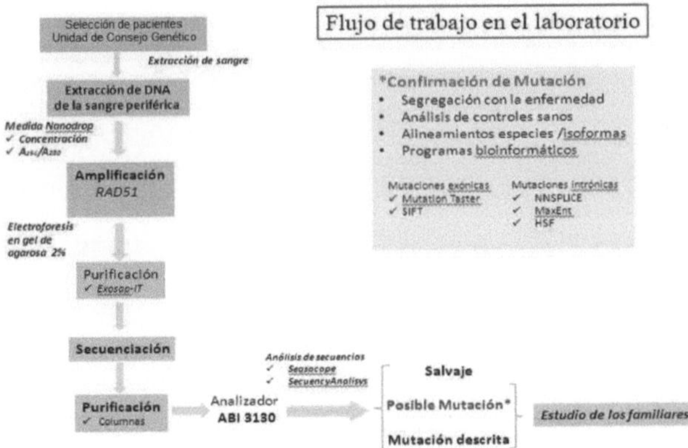

Figure 6: *Workflow in the LDG*

Distribution of tasks

The experiments for the molecular and bioinformatics studies were coordinated by the head of the Genetic Diagnostic Laboratory of the HCUVA.

The review of the available clinical information of the cases was collected by the oncology and clinical analysis physicians.

From the beginning of the study, patients were recruited in the corresponding section who met the inclusion criteria and who signed the informed consent form.   This process was carried out in the Hereditary Cancer Genetic Counseling Unit assisted by the physicians of the Medical Oncology Service of that hospital.

Patient inclusion review was conducted at the multidisciplinary breast and ovarian cancer committee meeting on a weekly basis.

After the preparation of the final report, the results were communicated at the meeting and the strategy to be followed was jointly assessed.

VIII. DISSEMINATION PLAN, APPLICABILITY AND PRACTICAL USEFULNESS

The applicability of the results in clinical practice stands out as it is a cross-sectional research project. The aim of the project is to provide knowledge for the clinical management of families with SCMOH that do not present variants in the *BRCA1* and *BRCA2* genes. The results were useful in the assessment of *RAD51C* and *RAD51D* as risk predictors in SCMOH syndrome and in the design of a global risk prediction model with other genes that allowed us to maximize the resources for the molecular diagnosis of mutations associated with SCMOH.

From the point of view of medical evidence, it is expected that the preliminary results will be the subject of communications to national and international congresses. The number of papers and the magnitude of the impact factor of the journals will depend on the size of the results.

IX. BIBLIOGRAPHY .

McPherson K, Steel CM, Dixon JM. ABC of breast diseases. Breast cancer-epidemiology, risk factors, and genetics. BMJ. September 9, 2000;321(7261):624-8.

Miki Y, Swensen J, Shattuck-Eidens D, Futreal PA, Harshman K, Tavtigian S, et al. A strong candidate for the breast and ovarian cancer susceptibility gene BRCA1. Science. Oct 7, 1994;266(5182):66-71.

Wooster R, Bignell G, Lancaster J, Swift S, Seal S, Mangion J, et al. Identification of the breast cancer susceptibility gene BRCA2. Nature. December 21, 1995;378(6559):789-92.

4. Antoniou AC, Hardy R, Walker L, Evans DG, Shenton A, Eeles R, et al. Predicting the likelihood of carrying a BRCA1 or BRCA2 mutation: validation of BOADICEA, BRCAPRO, IBIS, Myriad and the Manchester scoring system using data from UK genetics clinics. J Med Genet. July 2008;45(7):425-31.

5. Ferlay J, Autier P, Boniol M, Heanue M, Colombet M, Boyle P. Estimates of the cancer incidence and mortality in Europe in 2006. Ann Oncol Off J Eur Soc Med Oncol ESMO. March 2007;18(3):581-92.

6. Graña Suárez B. Hereditary breast/ovarian cancer syndrome: development of a clinical guideline and analysis of ATM, TP53 genes in high-risk families for breast cancer not associated with BRCA1 and/or BRCA2. February 11, 2013 [cited March 31, 2013]; Retrieved from: http://dspace.usc.es/handle/10347/7279

7. Ferlay J, Steliarova-Foucher E, Lortet-Tieulent J, Rosso S, Coebergh JWW, Comber H, et al. Cancer incidence and mortality patterns in Europe: Estimates for 40 countries in 2012. Eur J Cancer Oxf Engl 1990. February 25, 2013;

8. Ford D, Easton DF, Stratton M, Narod S, Goldgar D, Devilee P, et al. Genetic heterogeneity and penetrance analysis of the BRCA1 and BRCA2 genes in breast cancer families. The Breast Cancer Linkage Consortium. Am J Hum Genet. March 1998;62(3):676-89.

9. Díez O, Gutiérrez-Enríquez S, Ramón y Cajal T. [Breast cancer susceptibility genes]. Med Clinica. March 4, 2006;126(8):304-10.

10. Hall,J.M., Lee,M.K., Newman,B., Morrow,J.E., Anderson,L.A., Huey,B., and King,M.C. (1990) Linkage of early-onset familial breast cancer to chromosome 17q21. Science, 250, 1684-1689.

11. Miki,Y., Swensen,J., Shattuck-Eidens,D., Futreal,P.A., Harshman,K., Tavtigian,S., Liu,Q., Cochran,C., Bennett,L.M., Ding,W., and . (1994) A strong candidate for the breast and ovarian cancer susceptibility gene BRCA1. Science, 266, 66-71.

12. Wooster,R., Bignell,G., Lancaster,J., Swift,S., Seal,S., Mangion,J., Collins,N., Gregory,S., Gumbs,C., and Micklem,G. (1995) Identification of the breast cancer susceptibility gene BRCA2. Nature, 378, 789-792.

13. Malkin D, Li FP, Strong LC, Fraumeni JF Jr, Nelson CE, Kim DH, et al. Germ line p53 mutations in a familial syndrome of breast cancer, sarcomas, and other neoplasms. Science. Nov 30, 1990;250(4985):1233-8.

14. Nelen MR, van Staveren WC, Peeters EA, Hassel MB, Gorlin RJ, Hamm H, et al. Germline mutations in the PTEN/MMAC1 gene in patients with Cowden disease. Hum Mol Genet. August 1997;6(8):1383-7.

15. Luis Robles et al. Hereditary Cancer II Edition.

16. Meindl A, Hellebrand H, Wiek C, Erven V, Wappenschmidt B, Niederacher D, et al. Germline mutations in breast and ovarian cancer pedigrees establish RAD51C as a human cancer susceptibility gene. Nat Genet. 2010;42(5):410-4.

17. Osorio A, Endt D, Fernández F, Eirich K, de La Hoya M, Schmutzler R, Caldés T, Meindl A, Schindler D, Benítez J. Humar Molecular Genetics 2012 1-10.

18. Meindl A, Ditsch N, Kast K, Rhiem K, Schmutzler RK. Hereditary breast and ovarian cancer: new genes, new treatments, new concepts. Dtsch Ärzteblatt Int. 2011;108(19):323-30.

19. Roy,R., Chun,J., and Powell,S.N. (2012) BRCA1 and BRCA2: different roles in a common pathway of genome protection. Nat.Rev.Cancer, 12, 68-78.

20. Marquis,S.T., Rajan,J.V., Wynshaw-Boris,A., Xu,J., Yin,G.Y., Abel,K.J., Weber,B.L., and Chodosh,L.A. (1995) The developmental pattern of Brca1 expression implies a role in differentiation of the breast and other tissues. Nat.Genet., 11, 17-26.

21. Meza,J.E., Brzovic,P.S., King,M.C., and Klevit,R.E. (1999) Mapping the functional domains of BRCA1. Interaction of the ring finger domains of BRCA1 and BARD1. J.Biol.Chem., 274, 5659-5665.

22. Huen,M.S., Sy,S.M., and Chen,J. (2010) BRCA1 and its toolbox for the maintenance of genome integrity. Nat.Rev.Mol.Cell Biol., 11, 138-148.

23. Zhang,F., Ma,J., Wu,J., Ye,L., Cai,H., Xia,B., and Yu,X. (2009) PALB2 links BRCA1 and BRCA2 in the DNA-damage response. Curr.Biol., 19, 524-529.

Fabbro,M., Savage,K., Hobson,K., Deans,A.J., Powell,S.N., McArthur,G.A., and Khanna,K.K. (2004) BRCA1-BARD1 complexes are required for p53Ser-15 phosphorylation and a G1/S arrest following ionizing radiation-induced DNA damage. J.Biol.Chem., 279, 31251-31258.

Yu,X., Chini,C.C., He,M., Mer,G., and Chen,J. (2003) The BRCT domain is a phospho-protein binding domain. Science, 302, 639-642.

26. Moynahan,M.E., Pierce,A.J., and Jasin,M. (2001) BRCA2 is required for homology-directed repair of chromosomal breaks. Mol.Cell, 7, 263-272.

27. Esashi,F., Christ,N., Gannon,J., Liu,Y., Hunt,T., Jasin,M., and West,S.C. (2005) CDK-dependent phosphorylation of BRCA2 as a regulatory mechanism for recombinational repair. Nature, 434, 598-604.

28. Baumann,P., Benson,F.E., and West,S.C. (1996) Human Rad51 protein promotes ATP-dependent homologous pairing and strand transfer reactions in vitro. Cell, 87, 757-766.

29. Meindl A, Hellebrand H, Wiek C, Erven V, Wappenschmidt B, Niederacher D, et al. Germline mutations in breast and ovarian cancer pedigrees establish RAD51C as a human cancer susceptibility gene. Nat Genet. May 2010;42(5):410-4.

Gabaldó X, Sarabia Mesguer MD, Alonso Romero JL, Marin Vera M, Marin Zafra P, Sánchez Henarejos P, Ruiz Espejo F. Novel BRCA1 deleterious mutation (c.1918C>T) in familial breast and ovarian cancer syndrome Who share a common ancestry. Familial Cancer 2014. DOI 10.1007/s 10689-014-9708-5.

31.   Rebbeck TR, Mitra N, Domchek SM, Wan F, Chuai S, Friebel TM, et al. Modification of ovarian cancer risk by BRCA1/2-interacting genes in a multicenter cohort of BRCA1/2 mutation carriers. Cancer Res. July 15, 2009;69(14):5801-10.

32. HGMD®homepage [Internet].Retrieved from http://www.hgmd.cf.ac.uk/ac/index.php.

33. Home - LOVD - An Open Source DNA variation database system [Internet]. Retrieved from: http://www.lovd.nl/3.0/home

MaxEntScan:scoresplice   [Internet].   Retrieved   from: http://genes.mit.edu/burgelab/maxent/Xmaxentscan_scoreseq.html

BDGP: Splice Site Prediction by Neural Network [Internet]. Retrieved from: http://www.fruitfly.org/seq_tools/splice.html

36. Human Splicing Finder - Version 2.4.1 [Internet]. Retrieved from: http://www.umd.be/HSF/

SIFT - Tool to predict nonsynonymous / missense variants [Internet]. Retrieved from: http://sift.bii.a-star.edu.sg/

MutationTaster [Internet]. Retrieved from: http://www.mutationtaster.org/

Buy your books fast and straightforward online - at one of world's fastest growing online book stores! Environmentally sound due to Print-on-Demand technologies.

Buy your books online at
www.morebooks.shop

Kaufen Sie Ihre Bücher schnell und unkompliziert online – auf einer der am schnellsten wachsenden Buchhandelsplattformen weltweit! Dank Print-On-Demand umwelt- und ressourcenschonend produziert.

Bücher schneller online kaufen
www.morebooks.shop

Printed by Books on Demand GmbH, Norderstedt / Germany